Forecasting the Weather

Alan Rodgers and Angella Streluk

Heinemann Library
Chicago, Illinois

Published by Heinemann Library,
an imprint of Reed Educational & Professional Publishing,
Chicago, Illinois
Customer Service 888-454-2279
Visit our website at www.heinemannlibrary.com

Designed by Storeybooks
Originated by Ambassador Litho Limited
Printed in Hong Kong/China

07 06 05 04 03
10 9 8 7 6 5 4 3

Library of Congress Cataloging-in-Publication Data

Rodgers, Alan, 1958-
  Forecasting the weather / Alan Rodgers and Angella Streluk.
     p. cm. -- (Measuring the weather)
Summary: An introduction to weather forecasting, describing some of the
methods and tools used.
Includes bibliographical references and index.
  ISBN 1-58810-687-X -- ISBN 1-40340-127-6 (pbk.)
  1. Weather forecasting--Juvenile literature. [1. Weather
forecasting.] I. Streluk, Angella, 1961- II. Title.
  QC995.43 .R64 2002
  551.63--dc21
                          2002004025

Acknowledgments
The author and publishers are grateful to the following for permission to reproduce copyright material:
Science Photo Library, pp. 4, 23, 24; National Med Office/Crown, p. 6; Trevor Clifford Photography, pp. 7, 9, 12, 13, 20, 25; FLPA, pg. 8; Bruce Coleman Collection, pp. 17, 18; Robert Harding Picture Library, pp. 19, 29; Fotomas Index, p. 22; Photodisc, p. 26; National Met Library/R.S. Sconer, p. 27; Camera Press, pg. 29.

Cover photographs reproduced with permission of Jeff Edwards and Science Photo Library.

Our thanks to Jacquie Syvret of the Met Office for her assistance during the preparation of this book.

Every effort has been made to contact copyright holders of any material reproduced in this book. Any omissions will be rectified in subsequent printings if notice is given to the publisher.

Some words are shown in bold, **like this.** You can find out what they mean by looking in the glossary.

Contents

Weather

Are you interested in the weather? Have you ever tried to guess what the weather will be like? It is actually very difficult to predict! In this book, you can find out how professionals try to forecast the weather. This book will also give you ideas about what you can try to forecast and how you should do it. In the other books in this series, you can find out more about each type of weather, and how you can use simple instruments to help measure the weather.

Forecasting the weather has always been an interesting thing to do. With some knowledge and simple equipment, you can make your own reasonable predictions about the weather. By watching the weather carefully, you can collect **data** and then see if there are any patterns in this data. If you collect similar data in the future, this might help you to tell what the coming weather is going to be like.

It is the change in weather in the different seasons that triggers when trees and other plants grow and then shed their leaves. Farmers grow food according to these seasons. Predicting unexpected weather can make sure crops are not ruined.

We have all seen weather forecasts on television or in newspapers. These forecasts use simple picture symbols to represent the weather, so that it is easy to figure out what they mean. But the symbols used by professional weather forecasters are very different. **Meteorologists** who study the weather around the world all use the same symbols so that they can share information.

Weather all over the world

Although people are most interested in the weather where they live, it is really a global interest. This is because the Sun affects all parts of the earth. The movement of the earth causes day and night, which cause changes in temperature. The tilt of the earth gives us seasons. This means that different parts of the earth are heated at different times. This heat, or the lack of it, produces the weather.

The study of climate is also very important. Climate is the long-term weather over a number of years. Monitoring the climate helps meteorologists to see if there are changes in things such as temperature over a long period of time.

Be careful!

Do not look directly at the Sun when studying the weather. Also, never take shelter under trees during a thunderstorm, as they may be hit by lightning.

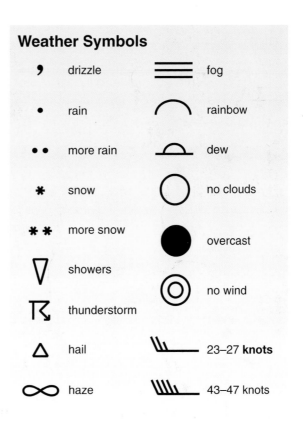

Weather Symbols

'	drizzle	☰	fog
•	rain	⌒	rainbow
• •	more rain		dew
✽	snow	◯	no clouds
✽ ✽	more snow	⬤	overcast
▽	showers	◎	no wind
⚡	thunderstorm		
△	hail		23–27 **knots**
∞	haze		43–47 knots

These are just some of the symbols used by professional meteorologists. They include symbols for cloud cover, **precipitation** (such as rain and snow), and wind speed.

What Is a Weather Station?

To record weather **data**, certain instruments are needed. These are often kept in the same location, and make up what is called a weather station. A weather station's main instruments are stored inside a special box called a Stevenson Screen. This was invented by Thomas Stevenson. He was the father of Robert Louis Stevenson, the author of the book *Treasure Island*. A Stevenson Screen is a white, rainproof box with a door. The sides are made from slats that allow air to move around the inside of the box, but that do not create drafts or let in rain. The door should face north if the weather station is north of the **equator**, or south if it is south of the equator. This means that when the door is opened, the Sun will not shine on the instruments and affect their readings. Smaller, less expensive Stevenson Screen kits are available for use by amateurs.

Several instruments are needed to record basic weather data. The best results will usually be obtained from the most expensive equipment that can be bought. However, good results can also be gained from quite cheap instruments.

Because a Stevenson Screen is white, it reflects the sunlight. The instruments for recording the weather will then not be affected by direct sunlight or drafts. This picture shows a Stevenson Screen with its door open.

Instruments for running a weather station

Thermometers are kept inside the Stevenson Screen. They are used for measuring temperatures. A **maximum and minimum thermometer** records the highest and lowest temperatures over a period of time. A **hygrometer** helps you figure out how much moisture **(humidity)** is in the air. A **weather vane** is used to show the wind direction. Rain gauges record how much rain has fallen. A **barometer,** which can be kept indoors, measures **air pressure.**

The next most useful addition to a weather station is a grass minimum thermometer, which is used to measure the temperature at ground level. You could use an ordinary maximum and minimum thermometer instead, placed just above some short grass near the Stevenson Screen. Another good instrument to have is a soil thermometer, which measures the temperature in the ground. If possible, have two of these, of different lengths.

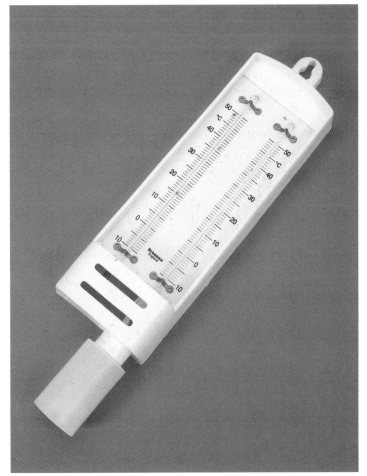

This hygrometer has two thermometers. One thermometer has its **bulb** wrapped in a wet **wick.** This thermometer always has its temperature reduced by the **evaporation** of the water in the wick. The readings from the two thermometers are compared to find out the **relative humidity.**

Where to Place a Weather Station

Meteorologists all position their weather instruments in a standard way so that **data** can be shared and compared. They give exact details about the location of their weather stations, including the height above sea level. This information helps meteorologists understand how data can be different depending on the location of a weather station.

Your weather station should be placed on level ground that is covered in short grass. It should be well away from large obstacles, like trees or walls. The Stevenson Screen should be set firmly on a stand, so that the instruments are at a height of 4 feet (1.2 meters) above the ground.

Measuring the weather

To find out the wind direction, you need a **weather vane.** It should be placed well away from obstacles that would prevent the wind from reaching the vane. It is best if the weather vane is two to four times the height of the nearest obstacle. To measure the strength of the wind, stand where you can see things blowing around. Then use the **Beaufort Scale** (see page 11) to estimate the wind's strength.

Weather instruments are kept close together so that their readings all come from the same location.

Clouds and rainfall

Knowing what types of clouds are in the sky will help with your forecast. Stand where you can clearly see the sky to look at a cloud type and the amount of this cloud in the sky. When you are measuring the weather, never look directly at the Sun. It can damage your eyes.

To measure rainfall, use a rain gauge. It should be located away from tall objects so that it is not sheltered from the rain. The rim of the funnel on the gauge should be 12 inches (305 millimeters) above the ground. This is so rain does not splash up from the surface and give false readings. The diameter of the funnel should be 5 inches (127 millimeters).

Thermometers

Some thermometers should be placed outside the Stevenson Screen. Grass minimum thermometers are used to measure the lowest temperature on the ground. They need to be set in an open space, just above the level of short grass. Soil thermometers measure the temperature below the surface of the ground. They should be set into the ground inside a hollow steel tube. Any snow that falls near them should be cleared away, because it can affect the thermometer's reading.

Grass minimum thermometers (like the one shown at the bottom of this picture) can be expensive, but they are very accurate instruments. An inexpensive alternative is an ordinary **maximum and minimum thermometer,** like the one shown at the top.

Recording Weather Data

It is important to collect weather **data** regularly. When you collect data regularly, you will start to notice patterns in the weather. Even with simple instruments, useful data can be collected. Don't worry if you do not have an ideal place to set up your instruments. Record the data on a sheet labeled with the date and the location from which the data was taken. If someone is interested in your data, this information will help them to understand it.

Try this yourself!

Here are some things to record when you are collecting weather data:

- Maximum temperature—the highest temperature for the previous 24 hours.
- Minimum temperature—the lowest temperature for the previous 24 hours.
- **Dry bulb temperature**—the current air temperature.
- **Wet bulb temperature**—check that there is water in the container that keeps the **wick** wet.
- **Relative humidity**—this can be calculated from a set of wet and dry bulb temperatures.
- **Air pressure**—read in millibars or kiloPascals from a **barometer.**
- Wind direction—which way the wind is blowing from.
- Wind strength—look at the trees and other movable items and assess the number on the **Beaufort Scale.**
- Cloud amount—record the amount of cloud cover by dividing the sky into eight or ten parts. If the sky is completely cloudy, your reading will be eight (or ten). If the clouds cover half of the sky, record it as four (or five).
- Cloud type—record the main cloud type for the sky. (See the Glossary for the names of some of the clouds.)
- Rain—record the contents of the rain gauge in inches or millimeters and then empty it.
- Snow—measure the depth in inches or millimeters with a metal ruler held vertically. Every day, brush clean an area so that you can measure fresh snowfall.
- Present weather—make a note of the present weather. Is it raining, snowing, or sunny?

The best time to record data is at 9:00 A.M., which is the time that many other weather stations record their data. Your data could then be compared to theirs.

Weather Data Collection Sheet

Date: Sept. 2002   Place Delaware, OH

Date	Day	Max. temp.	Min. temp.	Dry temp.	Wet temp.	Relative humidity	Pressure (millibars)	Wind direction	Wind strength	Cloud amount	Cloud type	Rain/ snow	Present
1	M	63.5° F	54.5 F	59° F	56.3° F	84%	995	S	3	6	Cumulus	2.5 in.	Wet!
2	T	64.4° F	54.5° F	57.2° F	55.4° F	89%	998	NW	2	8	A.stratus	1 in.	Showers
3	Wed	62.6° F	50.3° F	55.4° F	53.6° F	88%	1007	NW	2	0	Cumulus	0.25 in.	Bright
4	Th	64.4° F	51.8° F	57.2° F	54.5° F	83%	1012	NW	2	7	S.cumulus	0 in.	Cloudy
5													

Make your own weather data collection sheet. Here is an example that you can copy and use if you like. You can easily produce data collection sheets if you have a computer.

The Beaufort Scale

0 calm	1 light air	2 light breeze
smoke rises vertically; water smooth — less than 1 mph (less than 1 kph)	smoke shows wind direction; water ruffled — 1–3 mph (1–5 kph)	leaves rustle; wind felt on face — 4–7 mph (6–11 kph)
3 gentle breeze loose paper blows around — 8–12 mph (12–19 kph)	**4 moderate breeze** branches sway — 13–18 mph (20–29 kph)	**5 fresh breeze** small trees sway; leaves blown off — 19–24 mph (30–39 kph)
6 strong breeze whistling in telephone wires; sea spray — 25–31 mph (40–50 kph)	**7 near gale** large trees sway — 32–38 mph (51–61 kph)	**8 gale** twigs break from trees — 39–46 mph (62–74 kph)
9 strong gale branches break from trees — 47–54 mph (75–87 kph)	**10 storm** trees uprooted; weak buildings collapse — 55–63 mph (88–102 kph)	**11 violent storm** widespread damage — 64–72 mph (103–116 kph)

12 hurricane
widespread structural damage
above 72 mph
(above 116 kph)

These drawings help you figure out the wind's strength according to the Beaufort Scale. Look around you and decide which one is right for the wind you feel. The wind speeds are in miles (kilometers) per hour.

Graphs and Charts Using ICT

Collecting **data** would be a waste of time if it were never looked at again. Using **ICT** (**Information and Communication Technology**) helps you to spot patterns in the weather more easily. You can type in data, edit it, and print out copies. Data can be sent to others via e-mail. It can also be put onto web pages so that people around the world can look at it.

You can use a computer to record and analyze your weather data to find patterns.

Just like a professional **meteorologist,** you can use a **spreadsheet** to create graphs from the data. Choose the type of graph carefully. For rainfall, use a bar chart to make it easy to see which day had the most rain, the least rain, or the same rain as another day. For temperature, use a line graph to show temperature increasing and decreasing. A pie chart can show which wind direction was most common.

Try this yourself!

Use a spreadsheet to make these calculations:

- Finding totals—for example, how much rain has fallen.
- Finding the sum—for example, how many days the wind blew from each direction.
- Finding averages—for example, the average dry bulb temperature.
- Finding maximum/minimum temperatures.

If you don't have a computer, fill in your data sheet by hand and use a calculator for the calculations.

Finding errors

If you put two or more sets of data into a graph, it will help show if you have made any mistakes. Wet and dry bulb temperatures can be graphed together. If the **wet bulb temperature** is higher than the **dry bulb temperature,** then data has been recorded or entered incorrectly. Maximum and minimum temperatures can be recorded on the same graph. If the minimum temperature goes above the maximum, something has gone wrong!

Once you have a spreadsheet with graphs, the file can be saved and reused month by month. Each month, type in the data and resave it using a new file name to help identify it.

Professional meteorologists have special **software** packages to help them present their reports. These packages can arrange weather icons on maps, and can include charts to explain the changes in the weather.

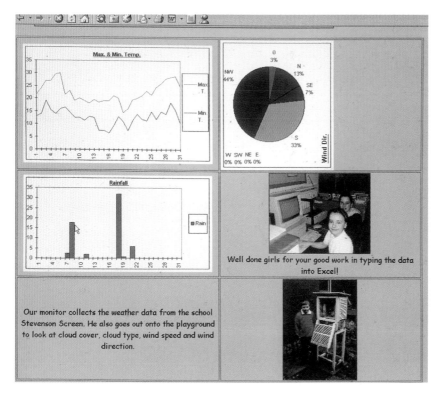

Well done girls for your good work in typing the data into Excel!

Our monitor collects the weather data from the school Stevenson Screen. He also goes out onto the playground to look at cloud cover, cloud type, wind speed and wind direction.

This spreadsheet, which shows graphs of the weather in a town in the U.K., is uploaded onto the Internet each month. People around the world can use it to see what the weather is like in that place.

Ideas for Forecasting Weather

Any amateur **meteorologist** would like to predict the weather. This is not a simple task! Professional weather forecasters are very good at making short-term forecasts. However, they use very sophisticated **data-**collecting devices and powerful computers. It is more difficult to make accurate long-term forecasts. This is because weather is so unpredictable! Even so, you can do a number of things to try and predict your weather, even if you don't have professional instruments.

Everybody wants to know if it will rain. This is not easy to forecast, but there are some pointers that will help you predict if there is going to be **precipitation.** Try using the chart below to predict precipitation. This chart refers to low-level clouds only. It also requires readings from a **barometer.**

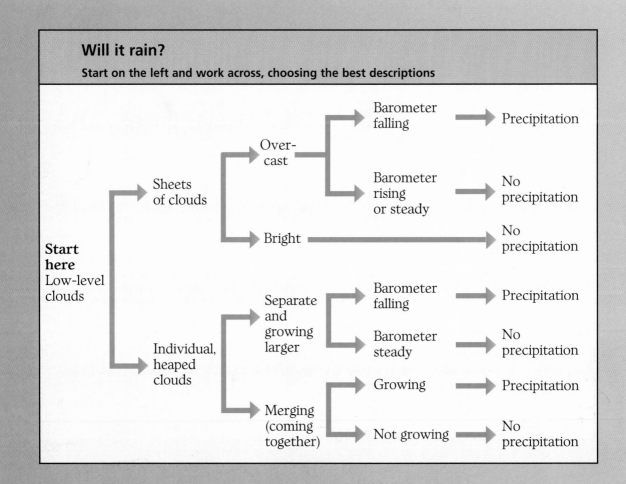

Will it rain?
Start on the left and work across, choosing the best descriptions

Making your own predictions

If you want to make your own weather predictions, you need to think about the things that will help you collect the right data. The chart below shows some of the factors that you must consider.

You can predict the overnight temperature to tell whether there will be a frost or not. The temperature usually falls at a steady rate during the night. This means that you can predict what the lowest temperature will be. Read the temperature at dusk, and then an hour and a half later. Make a graph showing the temperatures and hours until dawn. Draw a straight line joining the two temperature readings. Continue the line with a ruler and see if it goes below freezing before dawn.

Keep good records of what weather you predict. Write down what you thought would happen and compare it with what did happen. Good ideas can be used again, and those that did not work can be changed.

Things to consider when predicting the weather			
Time of day	Midnight	Dawn	Midday	Dusk
Season	Winter	Spring	Summer	Fall
Temperature	Very cold	Cold	Warm	Hot
Air pressure	Very low	Low	High	Very high
Air pressure trend	Steady	Falling	Rising	Erratic (changing)
Wind strength	Calm	Light	High	Very high
Wind direction	North	East	South	West
Humidity	Low	Medium	High	Very high
Cloud amount	Clear	$\frac{1}{4}$ of sky	$\frac{1}{2}$ of sky	Full sky
Cloud type	**Cirrus** (feathered clouds)	**Cumulus** (heaped clouds)	**Stratus** (layered clouds)	Giant clouds
Precipitation	None	Drizzle	Rain	Snow

Making a Local Weather Forecast

Weather can be a very local event. It sometimes rains at the front of a house and is dry at the back! The better you know your local area, the more likely it is that you will be able to predict what will happen to the weather there. If you live near water, there will be a lot more moisture in the **atmosphere.** You could find out more about this by keeping a record of wind direction and rainfall. You should soon be able to tell which winds bring wet weather to your area, and which bring dry weather.

If you have to guess what the weather will be without observation or using instruments, guess that the weather tomorrow will be the same as it was today. There is a seven in ten chance that you will be right!

Where you live

The geography of the area in which you live will influence the type of weather you should expect. You will not have sea showers or sea breezes if you live more than 30 miles (48 kilometers) from the coast. Large hills of several hundred feet or more can produce different kinds of weather in the local area. The side of the hill that the wind is blowing into will receive more **precipitation** than the opposite side.

The chart below shows some ways in which geography can affect the weather. The date and time can also make a difference to the weather you can expect. The chart will not help you to predict the weather on a specific day, but will give you some things to think about, especially when you visit different places.

Ways in which geography can affect the weather					
Location / Time	Seaside	Inland	Near water	Mountain	Valley
Day	Sea breeze	Warm	Cool	Cool	Warm
Night	Land breeze	Cool	Fog	Frost	Fog
Winter	Storms	Cold	Cool/**lake effect**	Very cold	Cool
Summer	Breezes	Hot	Warm/showers	Cool	Warm

Try this yourself!

Predict whether a **cumulonimbus** cloud will bring a storm.

- Is it raining heavily?
- Is the **air pressure** falling?
- Is the wind getting stronger?
- Is the **humidity** rising?
- Is there a giant cloud with an **anvil** head at the top?

If the answer to all those questions is yes, then it is likely that a storm is approaching.

When a cumulonimbus cloud approaches, it's time to go indoors and watch the rain pour down. You may also hear thunder and see lightning. This cloud may even bring hail.

Weather Sayings

For thousands of years, people have relied on farmers to provide food. For most of that time, there were no professional weather forecasts. As farming is dependent on the weather, people used sayings to pass on ideas about forecasting the weather. However, people did not keep records of the success of these sayings, so some may be more useful than others!

Different meanings

Some weather sayings have more than one meaning. Perhaps people from different areas interpreted them differently. Some sayings are not very reliable!

"Cows lie down before rain."
Some people believe that if cows lie down, it is going to rain soon. Others believe that if cows lie down, it is a sign of good weather to come. It is obviously not a very trustworthy saying!

Is this cow lying down because rain is coming, or because nice weather is coming? Cows probably do not take much notice of the showers, and are not a very good guide to the weather!

However, there are some interesting sayings about the weather and forecasts that have some truth in them. Here are some of them:

"Rain before seven, fine before eleven."
Rain brought by a **frontal system** often lasts less than four hours as it passes overhead, so this saying is usually true. This saying is good for a forecast at any time of day, not just before 7:00 A.M.!

"Red sky at night, sailor's delight,
Red sky in the morning, sailors take warning."
Red skies in the evening are caused by dust in the dry western sky. This means that there are no clouds to bring rain. If there is a red sky in the morning with layered clouds that are lit up by the Sun, it can mean that the dry weather has passed and there may be some rain on the way.

"Mackerel sky, mackerel sky, never long wet, never long dry."
A **mackerel** sky has a sheet of clouds in rounded heaps. This means that rain will arrive sometime between six hours and three days later. Generally, if the mackerel sky appears suddenly out of a clear, blue sky, the rain will arrive sooner rather than later.

It is not surprising that a weather saying was based on this beautiful sky. Red sky at night only works for the northern **hemisphere,** where weather generally moves from west to east.

Professional Weather Forecasts

Radio, television, newspapers, and the Internet all give us lots of **data** and information about the weather.

To make use of professional forecasts, you need to know roughly where you live on a map of your country. Your own location will not always be marked on the map, so you must look for the cities that are nearest to you to help figure out exactly where your area is. You will also need to know the meanings of the symbols that **meteorologists** use. These are usually easy to understand. They represent clouds, **precipitation,** temperature, and wind and **air pressure.**

In detailed television forecasts, two types of weather maps are usually used. The first are large-scale maps of an area with lines called **isobars,** which show air pressure and how the weather is moving. The second kind of map has symbols that look like drawings. These show details of wind, precipitation, and temperature. Television and radio forecasts are also used to give warnings about severe weather.

Television weather reporters provide useful weather data in a way that can be easily understood. The presenter describes the weather situation and points to the symbols to help explain the situation. The presenter also gives forecasts showing how the weather will change over the next day or so.

The weather reports in newspapers also use symbols that are easy to understand. They often include weather facts about the region, country, and world. They also give details of information important to the whole country, such as the **Heat Index (HI)**. The Heat Index shows how hot it feels according to temperature and **humidity.**

For up-to-date weather information, the Internet is useful. You can learn about current weather and forecasts in almost every place in the world.

Forecasts for weather at sea

Weather forecasts apply to the oceans as well as the land. Oceans are divided into areas so that ships can listen to the weather forecast for whichever area they happen to be going through. These shipping forecasts are read on the radio at the same times each day. Details are always read in the same order, so that people at sea will know which information is coming next.

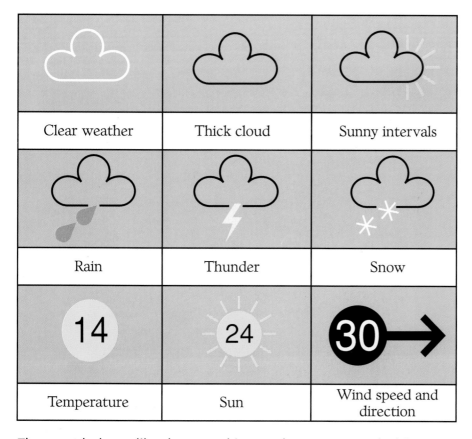

Clear weather	Thick cloud	Sunny intervals
Rain	Thunder	Snow
Temperature	Sun	Wind speed and direction

These symbols are like those used in weather maps on television and in newspapers. They vary in style, but all give the same basic information.

How the Professionals Do It

Weather watchers work together a lot. Every day, thousands of weather stations around the world share their **data.** This means that it is almost possible to show what the whole world's weather is like.

The idea of sharing weather data has developed over the years. By 1849, daily weather maps were displayed to the public in the **Smithsonian Institution** in Washington, D.C. At the Great Exhibition of 1851, held in London, daily weather reports were displayed for the public to read. In the 1850s, the British government founded the first real Meteorological Department. When the first forecasts were given to the public, there was much discussion about how accurate they were.

Mathematical weather forecasts

Early weather forecasters thought that the earth could be divided up into sections, and that observations could be made in each section. These observations could then be made into a set of numbers, which would give a weather forecast. In 1922 Lewis Richardson published this idea in his book, *Weather Prediction by Numerical Process.* He imagined that thousands of mathematicians could be used to figure out a forecast. His idea had to wait until the invention of high-speed computers in the 1950s.

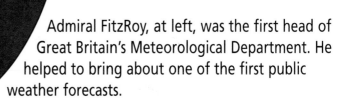

Admiral FitzRoy, at left, was the first head of Great Britain's Meteorological Department. He helped to bring about one of the first public weather forecasts.

The idea of predicting the weather is very complicated. The amount of data needed is huge, as there are so many things that affect the weather. The most powerful weather computers in the world work around the clock in Washington, D.C., and Bracknell, U.K. The data collected from around the world is turned into digital data that can be processed by these computers.

Once the data is processed, experienced **meteorologists** do the final work on the forecast. They have the latest data from **satellites,** and can spot any errors made by the computer. While the forecasts are being completed, the next batch of data is already being processed.

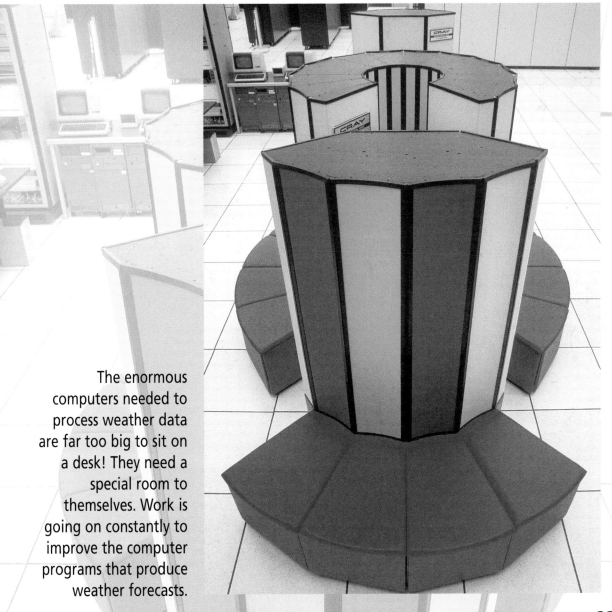

The enormous computers needed to process weather data are far too big to sit on a desk! They need a special room to themselves. Work is going on constantly to improve the computer programs that produce weather forecasts.

Finding Out about Weather around the World

Before the 1800s, famous people like Thomas Jefferson, George Washington, and Benjamin Franklin were interested weather watchers. They encouraged other people to be interested in the weather.

In the 1830s, Samuel Morse invented one of the first fast communication services: the telegraph system. This meant that people could exchange weather **data** much more quickly. By 1849, daily weather maps were displayed in the **Smithsonian Institution** in Washington, D.C. After ships were lost in storms in the Crimean War (1853–56), the British government set up the English National Weather Service in 1854. The American service was set up after storms in the Great Lakes in 1868 and 1869. By 1873, an international meteorological organization was established, and in 1950, the present World Meteorological Organization was founded. Today, this organization helps thousands of people and governments to measure the weather. They try to turn weather data into a pattern of weather predictions for the whole world.

This satellite image shows the white swirls of the clouds above the earth. It can be put onto the Internet very soon after it is received from the satellite.

Weather satellites

A lot of data comes from **satellites** orbiting about 22,000 miles (36,000 kilometers) above the earth. A series of five satellites can cover the middle section of the earth. They produce pictures of the part of the earth they can see. The other satellites used are the **polar**-orbiting satellites. These travel around the earth at a height of about 560 miles (900 kilometers) They can only see a section of the earth at a time, but they see it in greater detail. They usually cross the **equator** about fifteen times a day on their orbits.

Satellites produce lots of pictures. Some show data as color-coded pictures. The images may be used to show thickness of clouds, temperature, or the height of waves. These complicated images help **meteorologists,** but are not so easy for the general public to understand. Other satellite images can be understood by everybody. Many universities and government groups display these simpler, up-to-date weather images on the Internet. These can now be used by weather enthusiasts as well as professional weather reporters.

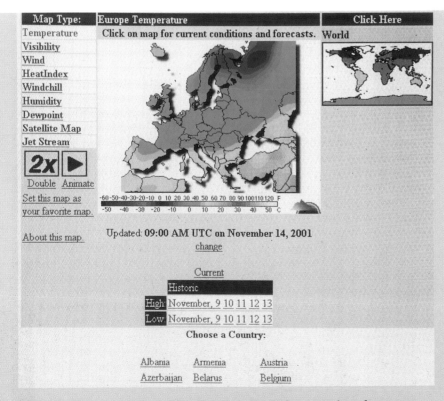

You can find a current weather report and a weather forecast for a specific location on the Internet.

Photographing the Weather

If you are interested in the weather, why not take photographs of interesting weather conditions? Don't worry if some of your photographs don't come out very well—photographing the weather can be difficult.

Whenever you are photographing the weather, you must follow a very important rule. Never look directly at the Sun or point your camera at the Sun. You will damage your eyesight.

The type of camera you use will make a difference in your final picture. Basic or disposable cameras will give interesting results, but a more sophisticated camera will give even better photographs. Color film is probably best for taking pictures of the weather, but some colors in the sky, such as purples and oranges, may not turn out as you expect. Record the key facts on the back of the photographs when they have been developed. Arranging them in an album will make them more interesting. You could sort them in the order you took them, or by theme (for example, different cloud types).

The patterns and textures in the weather can be very interesting. Frost (shown here) and dew can make stunning subjects for photography.

Digital pictures

If you have a **digital camera,** you can take photographs and see what they look like without having to get them processed. It is possible to take a photograph, see what it looks like, and then retake it if necessary. You can also edit the images and change their colors and brightness with **ICT software.** Digital images can easily be made into **animations.** For example, an image of the same cloud can be taken every few minutes. These can then be displayed quickly one after another, showing the changes in the cloud's shape. This animation will look similar to a video clip. Digital images can easily be displayed on web sites.

Try this yourself!

Use a notebook to record the key facts about each picture you take.
- What was photographed?
- Where was it taken?
- When was it taken (the date and time)?
- Why did you think it was interesting?
- What camera settings, filters, or lenses were used?

These photographs were taken at regular times, and could be made into an animation to show the movement or formation of clouds.

Record-Breaking Weather

Many people are interested in records and record-breaking events. Weather records can be very interesting. Can you remember the hottest day in your area? The table below compares **data** from towns and cities across the world, all called Tamworth. Remember that a day that is hot for Tamworth in the U.K. could be thought of as a quite cool day for the people of Tamworth, Australia!

Why do *you* think that people are so interested in record breakers?

Weather data for the warmest months in each location				
	Average high temperature	Average low temperature	Average precipitation	Month
Tamworth, New South Wales, Australia	90°F	63°F	3.6 inches	January
Tamworth, Staffordshire, U.K.	71°F	55°F	2.2 inches	July
Tamworth, New Hampshire	79.5°F	52°F	4 inches	July
Tamworth, Virginia	88°F	67.5°F	5 inches	July

This data uses the nearest available weather data for each place. Not all of the weather stations are official sites. The data collected was in different units of measurement. These had to be converted so that they could be compared. To compare summer temperatures in the two **hemispheres**, different months had to be used for the data.

The four towns around the world called Tamworth have very different weather. It is better to compare their averages—their usual weather—than simply to say which one is hottest or wettest by looking at their extreme weather. You will then find out more about the weather conditions in these places.

Tamworth, N.H.

Tamworth, Va.

Tamworth, Staffordshire, United Kingdom

Tamworth, New South Wales, Australia

One of the best ways of finding out current records is by looking on the Internet. Remember to check very carefully exactly what the record says. Use your knowledge to figure out what the records mean. Are the record-breakers telling the whole story? Are they claiming a world record, or one just for their area? You never know—if you measure the weather, one day you might measure a record!

World records
Records are fun, but they say more about the unusual than the usual. Here are some extreme world records:
- Highest world temperature: 136.4 °F (58° C) in Al Azizia, Libya, on September 13, 1922.
- Lowest world temperature: -128.56 °F (-89.2° C) in Vostok, Antarctica, on July 21, 1983.
- Highest world annual total rainfall: 1,041.8 inches (26,461 millimeters) in Cherrapunji, India, from August 1860 to July 1861.
- Highest world daily rainfall: 73.6 inches (1,870 millimeters) in Cilaos, La Réunion, on March 15–16, 1952.

The places holding the world records for the highest and the lowest ever recorded temperatures are never likely to have similar weather!

Glossary

air pressure pressure, at the surface of the earth, caused by the weight of the air in the atmosphere

animation series of images that are displayed quickly, one after the other, to give the impression of a moving picture

anvil iron block on which metal is hammered into shape. In this book, it is part of a cloud that is this shape.

atmosphere gases that surround the earth

barometer instrument for measuring air pressure. It may also show whether air pressure is rising or falling.

Beaufort Scale system of recording wind speed, devised by Francis Beaufort in 1805. It is a numerical scale ranging from 0 to 12; calm is indicated by 0 and a hurricane by 12.

bulb rounded end of the glass tube of a thermometer that contains the measuring liquid

cirrus highest form of clouds. They are made up of ice crystals in thin, featherlike shapes.

cumulus kind of cloud consisting of rounded heaps with a darker horizontal base

cumulonimbus cumulus cloud of great height. It often produces rain showers or is a sign of approaching thunderstorms.

data group of facts that can be investigated to get information

digital camera camera that records the picture in a form which can be understood by computers. It does not need film.

dry bulb temperature current air temperature. This can be used with the wet bulb temperature to figure out how much moisture is in the air.

equator imaginary horizontal line around the earth that divides the northern hemisphere from the southern hemisphere

evaporation process by which a liquid turns into a gas. In the weather, it usually refers to water turning into water vapor.

frontal system forward-facing edge between two air masses of different density and temperature

Heat Index (HI) method of calculating how hot it feels according to temperature and humidity

hemisphere half of a sphere. In geography, the world is split into two halves—the northern hemisphere and the southern hemisphere. The equator separates the two hemispheres.

humidity measurement of how much water vapor is in the air

hygrometer instrument for figuring out relative humidity

ICT (Information and Communication Technology) storing, processing, and presenting information by electronic means. This often involves using a computer.

isobar curved line on a weather map linking points of equal air pressure

knot nautical mile per hour, used in navigation and meteorology. One nautical mile is equal to 6,080 feet.

lake effect weather effect that causes large amounts of precipitation to fall on places near lakes or other large areas of water

mackerel type of fish. In weather terms, a "mackerel" sky is made up of lots of clouds that look like the pattern on a mackerel's back.

maximum and minimum thermometer thermometer that shows both the highest and lowest temperatures reached during a certain time

meteorologist person who collects weather data and studies weather

polar having to do with either the North or the South Pole

precipitation moisture that falls from clouds in a variety of forms, including rain, snow, or hail

relative humidity meteorologists' term for humidity. It is usually given as a percentage.

satellite man-made device that orbits around the earth, receiving and transmitting information

Smithsonian Institution group of museums and galleries in Washington, D.C., that are dedicated to public research and learning

software computer program or programs

spreadsheet computer program for processing data. The data is often in the form of numbers.

stratus layered cloud

weather vane piece of equipment with a pointer that shows the direction from which the wind is coming. It is labeled with the points of the compass.

wet bulb temperature temperature read with a special thermometer that is kept wet by a wick. The reading is used to figure out the amount of water in the air.

wick woven piece of cloth that sucks up a liquid

More Books to Read

Breen, Mark and Kathleen Friestad. *The Kids' Book of Weather Forecasting: Build a Weather Station, "Read" the Sky and Make Predictions!* Charlotte, Vt.: Williamson Publishing Company, 2000.

Kahl, Jonathon D. *Weather Watch: Forecasting the Weather.* Minneapolis, Minn.: Lerner Publications, 1996.

Scholastic, Inc. Staff. *Wind and Weather: Climates, Clouds, Snow, Tornadoes and How Weather is Predicted.* N.Y.: Scholastic, Inc., 1995.

Index